SOTHIS PRESS

Sothis Press aims to completely reimagine the public domain book with an emphasis on universal access to captivatingly designed print editions. We believe that technology will enable every single book ever published to be perpetually accessible in beautiful print form. Sothis is beauty in ubiquity.

www.sothis.press
1 Waterhouse Square, London, ECIN 2ST

Copper

From Mine to Finished Product

Anaconda Copper
Mining Company

SOTHIS PRESS

ISBN 978-1-0365-0705-3

This edition is published by
Sothis Press © 2023. The company
is registered in England and Wales.
The Company Number is 14900389.

ANACONDA AND THE UPPER WORKS IN 1887

COPPER

FROM

MINE TO FINISHED PRODUCT

ANACONDA COPPER MINING COMPANY

UNITED METALS SELLING COMPANY

SALES AGENTS

42 BROADWAY, NEW YORK

Hoists, Ore Bins and Timber Yards at the Leonard Mine, Butte, Montana

... mine development still practiced in the search for new ores in the Butte Distric

THE DEVELOPMENT OF A GREAT INDUSTRY

THE ANACONDA COPPER MINING COMPANY

In 1864, the same year that Montana was organized as a territory, gold was discovered on the hills of what is now the city of Butte, and in the next year there was considerable placer mining activity; but the gold was far from being so abundant as in many other gulches, and the prospectors soon turned to the silver-bearing quartz veins. In a few years Butte was a prosperous silver mining camp and several well-equipped mills were extracting this metal from the ore. Copper had been noticed from the first, yet very little interest was taken in it because the selling price of the base metals would not pay the high cost of transportation from a part of the country which as yet had no railroads. It was not until 1875 that any very systematic development of copper mines took place here, though there had been a small production before that date. In 1880 came a marked increase of production, amounting to a total of a million pounds in that year; but active copper mining did not really begin till 1882, when the yield was nine million pounds, the first railroad connection with the rest of the world having been completed in the preceding December. Since then the relative importance of copper, silver and gold have been exactly the reverse of what they were in the pioneer days of the state, the yearly supply from the mines of the district now amounting to hundreds of millions of pounds of copper, alloyed with small amounts of silver and gold. These facts, which correspond to the history of other mining regions, justify the proverb that "a man is likely to go to the poor house with a gold mine on his hands, may make a fair living from a silver mine, but easily becomes wealthy when he finds a good copper mine."

It is interesting and significant that the year 1880, which marked the beginning of Montana's important copper output, saw also the commencement of great progress in electrical engineering with its need for very large supplies of copper. While the industry has made possible the marvelous electrical developments of the present day, it has in turn benefited by the use of electric power machinery and the electrolytic method of refining. It is of special local interest that the ore is brought from Butte to Anaconda on a railroad that has been electrified for several years, as has also a large part of one of the transcontinental lines that serves the State.

We have said that the expensive transportation by pioneer methods delayed the development of copper mining; but there was also a geological reason why silver was in earlier favor: the action of the air and of percolating water had altered and leached out much of the copper from the upper levels, while leaving the silver. However, when the first mining shafts had gone through two or three hundred feet, vertically, of partly oxidized ground, the deposits were found to be rich in copper; the conditions became more and more favorable for marketing the red metal; and the production has increased so much that, since 1886, the Butte mines have annually yielded more copper than any other single district in the world.

Zinc is also found in these mines; and since 1910 there has been an increasingly large production of this useful metal. During the last quarter of a century most of the world's silver has come as a by-product of copper and lead smelting; and for many years Butte has been one of its foremost sources, because of the immense quantities of ore that are treated, although the precious white metal does not average more than two ounces per ton of ore. While the war was in progress considerable manganese ore, also, was shipped for use in making steel.

For nearly forty years the Anaconda Copper Mining Company has been the chief producer in the district; and its equipment has necessarily corresponded to its output. For a long time the city of Swansea, in Wales, was the world's principal center of copper smelting, and in the early eighties it still held

Operating an Air Drill—the first stage in breaking ore—Butte Mines

an important place, though its business in this line was declining. From 1882 to 1884, thirty-seven thousand tons of ore that averaged 45% copper was shipped to Swansea by the Anaconda Company; but in September of the latter year it began to treat its ore locally, in a plant that handled five hundred tons in twenty-four hours, and this made it unnecessary to continue selecting the very rich portions. The ore sent to the Works for the first few years had an average copper content of 12%. This was mechanically concentrated and then smelted into matte, the latter containing 64% copper. The matte was sent to Swansea and to Baltimore, for several years; but in 1892 a full installation of converters was provided to treat it, and since then nothing but metallic copper has been shipped. The size and capacity of plant have steadily increased with the growing demand for the red metal; and economies in production have permitted the handling of lower and lower grade ores. Those that are now treated contain 3% copper, which is less than in some of the waste products of the original plant. The present daily capacity is seventeen thousand tons of ore, or thirty-four times as much as in 1884; and the growth has been fairly regular through all these years.

An abundant supply of commercially important minerals, combined with skillful development of mining, metallurgical and business operations, has made the Anaconda Reduction Works the largest in the world among establishments producing any of the metals except iron and steel. The normal output is 16% of the copper produced in the United States, and more than 9% of the world's production.

The Works were located at Anaconda, twenty-six miles from the Butte mines, because there was no adequate water supply at any nearer point. The first plant was built on the north side of Warm Springs valley; the present one, which began operations in February, 1902, is on the south side of the valley, and is situated a mile east of the residential limits, on a hill-slope. Such a site, suitably graded for inter-department

An Underground Station at one of the mine shafts—Butte Mines

transportation, leads to great simplicity and convenience in the movement of all materials.

The Ores

There are three types of ore in the Butte mines, the first being chiefly valuable for its copper, the second for zinc, and the last for manganese. The main or central zone, which supplies three-quarters of the copper ore, has been the principal scene of mining operations in the district; but during the past ten years there has been increased activity in the intermediate zone, which surrounds the central area, and which is the source of most of the zinc ore. Mining for zinc has, indeed, been the main cause of this new activity. The intermediate zone also supplies a certain amount of copper ore; and manganese minerals begin to appear here. The presence of a good deal of manganese in the third or outer zone has long been known; but not until 1918 did market conditions favor the mining of manganese ore. Zinc is also important in this zone, but copper is very scarce.

The two principal copper minerals are chalcocite or glance (a sulphide of copper) and enargite (copper-arsenic sulphide), associated with pyrite (iron sulphide) in a gangue of quartz and altered granite. Bornite (a copper-iron sulphide) is present in small amounts; and covellite, the beautiful indigo-colored copper sulphide, is found but is not common. Ore is frequently mined that has the appearance of being mostly pyrite but which contains the above named copper minerals in finely distributed particles. Silver occurs in the central zone in the proportion of six-tenths of an ounce per ton for each per cent of copper; but the relatively small quantity of copper ore taken from the intermediate zone carries double this proportion of silver. Gold occurs only to the value of fifteen cents per ton of ore, but this is of some consequence considering the enormous tonnage treated.

Zinc occurs as sphalerite (the sulphide) associated with pyrite, rhodonite (manganese silicate) and rhodochrosite (the delicate pink manganese carbonate). These ores contain four-tenths of an ounce of silver per ton for

Electric Haulage of Ore on the 1,500-foot Level in the Tramway Mine

each per cent of zinc. Galena, the sulphide of lead, is found very sparingly.

The manganese is mostly in the form of rhodochrosite, but the oxide, pyrolusite, has also been mined in small quantities. There is a great deal of rhodonite in the outside zone, but this has not been mined because it is not acceptable to the steel makers. From this zone small quantities of silver ore are shipped, containing from ten to thirty ounces of silver per ton.

The various minerals that have been named are associated with much larger quantities of quartz and altered granite. Indeed, two-thirds of the material taken from the central zone is altered granite which, however, contains enough of the copper minerals in seams and veinlets to constitute ore. The ore from the intermediate zone contains 40% granite, with less pyrite but more quartz than occurs in the central area. The gangue in the outer zone is nearly all quartz.

Mining

Some idea as to the extent of mining operations can be formed when we say that, during the past forty years, the district has produced a hundred million tons of copper ore. A single train to transport this ore would stretch nearly half way round the globe. The copper content has amounted to eight billion (8,000,000,000) pounds; and if this were all drawn into telephone wire it would have the almost unbelievable length of fifty million miles, enough to put a continuous spiral girdle round about the earth, from pole to pole, with the successive turns but four miles apart.

Included in the Anaconda ownership there are twenty-nine main shafts and several auxiliary or air shafts. A few of the more important have reached the 3,400 foot level below the surface; and many of them are connected by elaborate drainage and ventilation tunnels on the 2,800 foot level. Much attention is given to ventilation, including the constant operation of immense power-driven fans. The timbering in a number of shafts has been fire-proofed with cement mortar, giving the appearance of concrete structures. Numerous drifts, or horizontal passages along the veins, are run out from these shafts at every hundred feet in depth; and these are connected by crosscuts and by vertical raises. Thirty-five miles of these passageways are driven each year; and their present total length is seven hundred miles. The volume of the stopes, from which the ore is actually extracted, is five or ten times the volume of the passages.

When an ore body of sufficient size and grade is encountered in one of the mines, some system is adopted that will permit it to be extracted profitably and with safety to the men employed. There are three general methods in use in the Butte mines, dependent on such conditions as width of vein, character of ore and solidity of the vein walls, namely, the square set, the rill-stopes and the back filling systems. About 80% of the mining is done by square setting, and nearly all the rest by rill-stoping. In any case five operations are involved: first the ground is drilled for blasting; then the holes are loaded with explosives and fired by means of a primer and fuse; next the ore is shoveled into chutes or cars; then the stope is timbered; and finally filled with waste rock, which is taken from development work in other parts of the mine.

The square set system consists of a structure composed of posts, caps and girts which are framed together and built into a substantial form of bracing. The stope advances in steps upward and along the vein. Wooden chutes are carried up with the stope, and provide means for conveying the broken ore from the stope floors to the haulage level. The rill-stope is a V-shaped excavation in the vein with its apex at the chute. The ore is broken from the sloping back and falls by gravity into the chute. This method uses less timber than the square set system, and requires much less shoveling, but is applicable only to certain veins.

The ore in the chutes is loaded into mine cars on the main levels and hauled to the stations by electric locomotives, where it is dumped into receiving pockets in the shaft. From these pockets it is loaded into skips and is hoisted to the surface where it is distributed in the main bins. Electrically operated ore trains then haul it from these bins to the Reduction Works.

Three Thousand Tons of Ore, and a Train of Box Cars Hauled by Electric Locomotives

Group of Structures—Anaconda Reduction Works

[illegible] Bins on the right. Small Cottrell installation on extreme right center, with searchlight tower on top of it. Main Stack on top of hill, No. 2 Roaster in center. Copper Concentrator shows, beyond the No. 2 Roaster, and [illegible] Building at lower extreme left. [illegible] Concentrator, we come to the Zinc Ore Crusher building, of brick; and below that, with stepped roof, is the main portion of

THE SMELTING OF COPPER

A BRIEF DESCRIPTION OF THE

ANACONDA REDUCTION WORKS

Outline of Processes

THE copper ores, delivered on the "high line" by the Butte, Anaconda & Pacific railway, are first enriched by mechanical concentration. In this process the useless part of the ore is separated from the valuable portion by making use of differences in specific gravity, the useless part being sluiced to the dump and 3% of copper in the ore becoming 8% in the concentrated product.

The product, which has been reduced to three-eighths of an inch and less in diameter, is roasted, to remove 80% of its sulphur, and is then smelted in reverberatory furnaces. When an extra large production of copper is called for, some of the ore, left in a coarser condition, is smelted in blast furnaces, in which the burning off of 60% of the sulphur and the melting of the charge are done in one operation. The reverberatories and blast furnaces produce matte with an assay of about 40% copper, and make slag as a waste product which is sluiced to the dump. The matte is treated in converters, which deliver an impure copper. The last stage is reached in the refining furnaces, which turn out the commercially pure metal; and this is shipped to the Company's electrolytic plants at Great Falls, Montana, and Perth Amboy, New Jersey, where the silver and the small amount of gold are separated. The metals are then marketed.

The zinc ore is increased from an assay of 12% zinc to 32% by mechanical concentration. The metallurgical treatment of this concentrate is conducted at the Company's plant in Great Falls, except as noted under the subject of roasting.

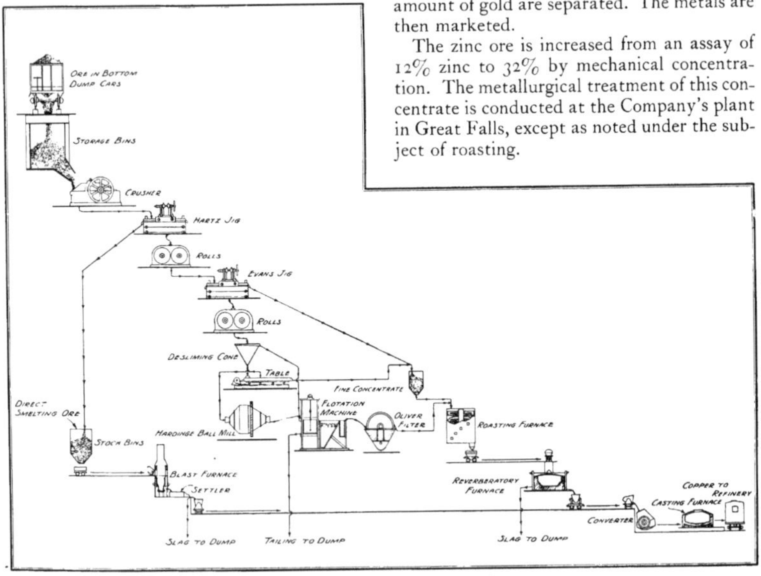

Flow of Material from Ore Cars to Copper Cars

Copper Concentrator

This department contains eight independent sections, each of which treats nearly twenty-one hundred tons of ore in twenty-four hours. One of these sections is here described.

The valuable minerals are separated in several stages, as much being taken out in the first stage as can be obtained of high enough value. The remaining part of the ore is then further crushed, to release the enclosed valuable minerals, and these in turn are separated as part of the concentrated product.

The ore as received from the mines ranges from fifteen inches in diameter to dust. The process begins by delivering the ore from the bin to a 12 by 24-inch Blake crusher, through two shaking feeders. The bottom plate of each feeder is perforated with two-inch round holes, through which some of the finer part of the ore drops, thus lessening the load on the crusher. The ore broken by the crusher passes to two trommels or revolving screens. These deliver the sizes that are still coarser than two inches, to two 8 by 20-inch crushers for further reduction in size, while the finer portion joins the undersize from the shaking feeders. All of the ore, now reduced to two inches and smaller, is raised to the top of the mill by bucket elevators, and is divided by a series of trommels for jigging, and a final undersize for other treatment. When the blast furnace department is in operation, the first two sizes, from two inches to three-eighths inch, are treated in water on Harz jigs, making concentrate, which is sent to the blast furnaces, and middling. The middling is ground finer in two sets of rolls, and sent back to the trommel system by way of the above mentioned elevators. When the blast furnaces are not in use, all of the two-inch to three-eighths inch ore is simply passed on through the rolls and the finer trommels and is treated on the machines farther on. The two sizes from three-eighths inch to one-sixteenth inch, are treated on Evans jigs, which also make concentrate and middling. This middling is ground to one-sixteenth inch and finer in rolls, and goes, together with that part of the original material that is finer than one-sixteenth inch, to Anaconda classifiers. In these, a rising current of fresh water causes the slime to overflow, while the sand discharges

Evans Jigs—Anaconda Reduction Works

Section of Hardinge Mills in Operation — Anaconda Reduction Works

from the spigots and passes to Wilfley tables. Here a further portion of concentrate is obtained, and the middling which remains is sent to Hardinge ball mills for still finer grinding. Each mill discharges the finely ground sand into a Dorr mechanical classifier. The latter sends back to the mill the portion that is not yet fine enough, while sizes from two one-hundredths of an inch to slime pass to the flotation machines. Here is the first point at which a waste product is made, tailing being washed to the dump from the flotation machines while another portion of concentrate is saved for smelting.

The slime that overflows from the classifiers preceding the Wilfley tables consists of 3% solid matter with 97% of water, and is thickened to 14% solids in circular tanks twenty-eight feet in diameter and three feet deep, placed in another building. Part of this thickened slime is returned to the main building for admixture with the finely ground sand as the latter enters the flotation machines, the proportions of sand to slime being eleven to 1. The remainder of the thickened slime is treated in another set of flotation machines in a special building. Here the final recovery of concentrate is made, and more tailing is sent to the dump.

It is of interest to compare the three different methods by which concentrate is separated from the ore. In each jig a plunger produces rapid upward pulsations of water through the bed of ore that rests on the sieve. These pulsations make the bed just loose enough for the valuable minerals to settle below the other parts of the ore, on account of their greater specific gravity. The layer of concentrate passes out through an adjustable trapped gate, while the middling flows over the end of the jig. The Wilfley tables move the thin bed of ore forward as the result of a peculiar jerk imparted by the driving mechanism; but the concentrate moves with greater momentum than the middling, due to its greater specific gravity. The shaking action slightly loosens the bed of ore, so that the concentrate settles below the middling. The wash water, flowing across the table at right angles to the direction of the jerking motion, carries the middling over the lower edge of the table while the riffles guide the

Flotation Machines—Anaconda Reduction Works

concentrate to the end of the table. In the flotation machine we have what may be aptly called an up-side-down process: the heavier minerals are made to rise in a mass of water while the lighter minerals sink. This is brought about by taking advantage of the fact that, when certain oils are mixed with the ore, they stick to the metallic minerals, but do not attach themselves to the non-metallic minerals. This selective action is aided by the use of small quantities of sulphuric acid. Vigorous agitation of the mixture of water, ore and oil beats air into the mass; and the presence of the oil makes a persistent froth of air bubbles, so light that it can support the fine mineral particles that adhere to the oil. The success of the flotation process depends on grinding the ore so fine that a considerable proportion of it is slimed or, in other words, reduced to the almost impalpable condition of clay. Particles even as fine as three one-thousandths of an inch in diameter give no satisfactory results unless mixed with the much finer, slimed ore.

The water is easily removed from the coarse sizes of concentrate by means of short stationary screens. From the medium sizes the excess water is gotten rid of in large settling tanks. The extremely fine, flotation concentrate also goes to settling tanks; but the settled product still contains 40% water. This is reduced to 15% by means of Oliver continuous filters, which are large revolving cylinders operating in steel tanks.

In the concentrator, 95% of the copper is saved by these relatively cheap processes, while nearly two-thirds of the ore is discarded. Previous to the introduction of flotation the recovery was low, for two reasons: earlier practice did not carry the fine grinding far enough to release all of the copper-bearing mineral, some of the tailing going to the dump in sizes as coarse as a sixteenth of an inch, which is three times the diameter of the coarsest particles now sent to waste; and the methods then in use for concentrating the very fine sizes were much less efficient than flotation.

Leaching Plant

Previous to adopting the flotation process of concentration, the tailing that was sent to the dump from the New Works concentrator carried an average of thirteen pounds of copper per ton, which is three times as much as is contained in the tailing that is produced now. Exposure to the weather and to the action of a small amount of carbonate of lime in the water continually running over the dump has converted 10% or 15% of this copper from sulphide to carbonate. The carbonate is difficult to recover by mechanical concentration, but is very easily soluble in acids. One of the recent improvements at the plant is the re-treatment of this old tailing by dissolving the contained copper in sulphuric acid.

By means of an electrically-operated excavator this old material is loaded into fifty-ton bottom-dump cars and is transferred to large storage bins at the leaching plant. From these bins the material is distributed to roasting furnaces in which the sulphide of copper is converted into oxide. These furnaces are described in the section on roasting. Their product is passed through rotary coolers to a mixer, where a little water is added to lay the dust; after which the material is delivered to immense leaching tanks by a series of rubber belt conveyors. The purpose in cooling the roasted material is to prevent rapid destruction of these belts.

The leaching equipment consists of fifteen redwood tanks, each fifty feet in diameter and fourteen feet deep, together with coils for heating, and pumps for circulating, the solutions. Five of the tanks are simply for storage of solutions, while the other ten are used for the actual extraction of the metal. These ten are lead lined. Cocoa matting is laid on the bottom, through which the solutions are filtered after they have dissolved the copper from the sand; and the matting is protected against excessive wear by wooden gratings laid over it.

A tank receives a thousand tons of roasted charge, which is distributed by a simple inclined movable chute, along the bottom of which there are several adjustable gates. The leaching is done by downward percolation of hot solutions containing suplhuric acid and salt. The main purpose of the salt is to dissolve silver, which would not be taken

Leaching Tanks, Showing Feed Distributors—Anaconda Reduction Works

up by the acid alone; but the salt also increases the speed of extraction of the copper. For each thousand-ton charge of sand there are used seventy thousand gallons of fresh water, thirty-five tons of commercial sulphuric acid and fifteen tons of common salt. The first liquor to be passed through a fresh charge of sand has already been used on previous charges, this being the standard method, in all such processes, in order to dissolve the maximum amount of metal in the minimum quantity of solution. After a suitable period of contact this first liquor is drawn off as the "strong solution" from which copper and silver are later precipitated. This solution is followed by others, successively weaker in copper, and, finally, by the fresh water. The spent charge is sluiced out of the tank through bottom gates, with water that has previously been used in the concentrator. This re-use of water is a necessary measure of economy which is practiced at various points about the Works. The total time required to charge, leach and discharge a tank is four and a half days.

Certain oxidized ores of copper are also leached. They do not need roasting, and have to be crushed only enough to make everything pass through a one-inch hole.

The copper and silver are recovered from the strong solution by passing the latter through a mass of scrap iron in a series of concrete tanks. The iron displaces the copper from solution and is itself largely dissolved, though part of the iron rusts and remains as rust with the so-called cement—or precipitated copper. This final product is dried and, as it then contains only 60% copper, is smelted in the reverberatory furnaces, which are described later.

Roasting

Roasting, in metallurgy, consists in burning the sulphide minerals to get rid of sulphur and to leave more or less of the metal content as oxides. The operation has to be conducted at a good red heat, yet below the temperature at which fusion would take place. In the case of copper smelting the aim is to remove only part of the sulphur, some of it being purposely left to form matte in the succeeding treatment; but in the case of zinc the sulphur is removed as completely as possible.

All of the roasting furnaces at this plant, whether for copper concentrate, zinc concentrate or the old tailing, are of the McDougall type. This is a cylindrical furnace, with several hearths one above another, substantially built of brick and enclosed in a steel-plate casing so that there shall be no injury to the furnace by expansion and contraction. The large amount of brick work absorbs a great deal of heat, while the cylindrical form, having less outside surface than any other shape, radiates less heat. Under these favorable conditions any fine ore that contains a large percentage of iron pyrite will roast without the use of other fuel, because the pyrite develops more heat than most minerals, its sulphur burning to sulphur dioxide gas and its iron to the solid, reddish brown oxide. When a furnace is being put into operation from a cold condition it must, of course, have a preliminary heating with wood or coal.

In number one roaster building the fine copper concentrate, from jigs and tables, is handled in sixty-four furnaces, each of which is fifteen feet inside diameter and has six hearths. The concentrate, automatically fed to the top hearth, is moved from the circumference to the center by two rabble arms which are attached to the central shaft, and drops through an opening to the next hearth, across which it is similarly moved to the circumference. The stirring action of the rabbles constantly exposes fresh surfaces of the charge to the hot gases which come up from the lower hearths. These gases dry out the 8% of moisture from the concentrate while passing over the top hearth; and before the charge has crossed the second hearth it is hot enough to begin roasting. The necessary air enters through the open doors of the bottom hearth. A dry sample of average feed contains 33% sulphur, which is reduced to 8% in the product. The latter is commonly called "calcine." Each of these furnaces has a capacity of forty-five tons, dry weight, of feed per twenty-four hours.

Number two roaster building serves especially for handling the flotation concentrate; but it also receives part of the table and fine jig products. There are twenty-eight furnaces in number two building, each having seven hearths and being twenty-three feet

Tops of Furnaces in No. 2 Roaster Building—Anaconda Reduction Works

inside diameter. Each of these furnaces handles one hundred and twenty-five tons, dry weight, of charge per twenty-four hours. The feed contains 12% moisture. The 30% of sulphur that remains after drying is reduced to 9% by roasting. Since the flotation concentrate is excessively fine, special attention was given to the detail design of these furnaces, in order to minimize the loss by dusting.

The furnaces in the leaching plant are similar to those in number one roaster building; but on account of the small percentage of sulphur in the feed, these furnaces receive extra heat from carefully regulated coal fires.

Some time after the zinc plant operations began at Great Falls, more roasting capacity was needed; and it was deemed better to use certain furnaces in the roaster building of the leaching department, here in Anaconda, than to build extra furnaces at Great Falls. For that reason, part of the zinc concentrate is roasted at this plant before shipping. This product contains too little pyrite for satisfactory self-roasting, so that coal fires are necessary. The 32% sulphur in the feed is reduced to 4% in the product; but three-fourths of this remaining sulphur is in the form of sulphate, which is readily soluble in the subsequent leaching process.

In all of these roasting furnaces the heat would warp the cast-iron rabble arms if suitable provision were not made for cooling. The arms are hollow, and water continually circulates through them, except in the case of the furnaces used to roast the old tailing. Air cooling is sufficient in that instance; and in cold weather the hot air from this source supplies most of the heat for the large building that contains the leaching tanks.

Anaconda Reduction Works

No. 1—Converter Building. No. 2—Coal Pulverizing Plant. No. 3—Acid Plant. No. 4—Reverberatory Furnace Building. No. 5—General View of Plant, Looking South. No. 6—Blast Furnace Building and Dust Chamber. No. 7—Brick Plant. No. 8—No. 2 Roaster Building.

Reverberatory Smelting

The principal part of a reverberatory furnace is a large chamber built of silica brick and thoroughly braced with steel buckstays. The material to be smelted is charged onto the hearth, and the heat that is needed to fuse it, is received by radiation from the incandescent flame of the fuel. Nine-tenths of the charge is calcine from the roasting furnaces, the rest being mostly dust produced in the various departments and caught in the system of chambers and flues that is described later. The charge contains a mixture of sulphides and oxides of copper and iron, together with silica, alumina and lime. A suitable amount of silica and alumina is purposely left in the product of the mechanical concentrating process to combine, during the smelting operation, with the oxide of iron that is produced in roasting. The product of this combination is slag; and by discarding the latter the copper remains in an enriched material—the matte. The lime is added, as finely broken limestone, in order to form a more easily fusible slag. To mix this lime with the rest of the charge, it is included as part of the feed to the roasting furnaces.

In the fused state copper has greater affinity for sulphur than have the other metals; consequently the copper oxide that is in the charge reacts with sulphide of iron to form copper sulphide, iron oxide and sulphur dioxide. This iron oxide, of course, becomes part of the slag and the sulphur dioxide passes away with the rest of the gases. Other similar reactions also eliminate sulphur, so that 4½% of the charge passes into the flue as sulphur dioxide and thus disposes of more than a quarter of the sulphur in the smelting mixture. Another 3½% of the charge goes off as carbonic acid gas from the decomposition of the limerock. The copper sulphide and the remaining iron sulphide melt together to form the matte, which has a specific gravity twice as high as the slag and hence settles to the bottom of the molten mass. From each hundred tons of charge, containing something over 9% of copper, there are produced twenty-four tons of matte, containing 38% of copper, and sixty-seven tons of slag. The latter runs off continuously into a stream of water which chills and granulates it and then transports it to the dump through a system of flumes. The matte is tapped from the furnaces periodically, and taken in immense steel ladles to the converters.

During the past few years reverberatory smelting has had marked attention from copper metallurgists, with corresponding improvements in practice. These improvements have been along three lines: first, a continuation of the striking increase in size of furnaces that has now been taking place for forty years; second, in the manner of charging the furnaces; and, finally, there has been much betterment in the method of using the large quantities of coal that are consumed.

The present furnaces have hearths twenty feet and four inches wide and one hundred and forty-three feet long; and the area is sixteen times as great as in the furnaces used by the Company in 1884. A hearth of the earlier date could, indeed, be put inside of the present structure with its long dimension across the width of the new furnace.

That early plant was laid out in such a way that it was necessary to quench the calcine with water before taking it to the matte furnaces, in order not to smother the men with the sulphur fumes in a blind tunnel into which the calcine was dropped from the roasters. This meant both a cold and a wet charge for smelting; and required a great deal of extra fuel to dry the charge and bring it back to the temperature that it had before wetting. At present as little time as possible is lost between roasting and smelting, and the charge enters the reverberatories at a temperature of eight hundred to nine hundred degrees Fahrenheit. In early practice the charge was shoveled into the furnace by hand, through side doors; but a great deal of cold air was, unfortunately, admitted at the same time. The present method is to drop the charge, by gravity, from overhead hoppers through iron pipes which are built into the roof of the furnace. This method admits but little cold air, requires far less time and less hard, disagreeable work, and permits heaping the charge against the sides of the furnace, thus protecting the walls and also exposing a larger surface of material to absorb the heat.

Coal Dust Firing at Reverberatory Furnace—Anaconda Reduction Works

Until a few years ago the coal was used in the time-honored way, on a grate at one end of the furnace; but the present standard practice is to grind the coal nearly as fine as flour and blow it directly into one end of the hearth with an air blast, where it burns like a gas flame, and the products of combustion pass away at the other end of the furnace. Since, by the new method, the combustion takes place wholly in the hearth itself, instead of partly on a grate that is separated from the furnace chamber by a bridge-wall, the heat is better utilized. Moreover, this method permits a nearly perfect control, so that there is a very uniform temperature condition and an almost constant quality of flame, which are decidedly more efficient than the unavoidably varying flame and temperature of the old-style firing.

All three of these changes have saved a great deal of fuel, which is the chief item of expense. Much of the heat from the fuel has also been saved in another way, which is not strictly a point in reverberatory operation, but rather one of general economy. This consists in passing the waste gases through a set of steam boilers. On account of the close contact of the gases with the boiler tubes, more heat is saved in the steam than is utilized in the smelting, which, as before indicated, receives most of its heat by radiation from the myriads of incandescent particles in the flame, as the latter passes over the charge, and only a little by direct contact. Though it would be desirable to utilize a larger part of the heat in actual smelting, yet it is seen what striking advance there has been in both efficiency and economy when we state that the smelting capacity of a single furnace has increased from twenty tons of charge per twenty-four hours, using ten or twelve tons of coal, in 1884, to five hundred and thirty tons of charge and eighty-five tons of coal in 1920, amounting to twenty-six times the capacity on sixteen times the hearth area with only eight times the tonnage of coal.

Blast Furnace Smelting

The blast furnaces are steel structures fifteen feet deep, the inside width varying from four and a half feet at the bottom to six feet at the top. There are two furnaces each fifteen feet long, one that is fifty-one feet long and another eighty-seven feet long, inside. The sides are water-jacketed as a protection against the intense heat developed by the smelting operation; and a constant flow of clear water is maintained through these jackets. Extending above this steel structure is a brick top as a convenience for disposing of the waste gases; and this upper portion has vertically sliding doors on each side just above the level of the charge floor, operated by means of compressed air.

Blast furnace smelting is best adapted to ore from a quarter of an inch to five or six inches in diameter; though more or less material outside of these limits is used. If fine material was treated, such as is sent to the reverberatories, the strong blast of air that passes through the charge would carry a large part of it out of the furnace and cause great losses. On the other hand, coarse ore cannot be economically treated in reverberatory furnaces, because the smelting would be too slow.

The smelting mixture contains the coarser portion of the product from the copper concentrator, coarsely broken limerock, some chilled slag that is returned from the converting process, and miscellaneous clean-up material from other departments. This mixture is brought from the stock bins in trains of special side-dumping cars. Under the spouts of the various stock bins there are automatic track-scales, an arrangement that permits proper proportioning of the different materials. When these cars arrive at the furnace they are dumped by hooking them up to compressed-air pistons. Just before putting each charge into the furnace, through the side doors, two hundred pounds of coke per ton of charge is introduced from large barrows. Air, supplied by great rotary blowers in the power house, at a pressure of forty-two ounces per square inch, is forced into the bottom of the furnace through numerous tuyere pipes; and, rising through the charge, burns the coke and also 60% of the sulphur content of the charge. The heat from the burning of this sulphur saves a great deal

Matte Settlers in Front of Blast Furnaces—Anaconda Reduction Works

Converters—Anaconda Reduction Works

of coke. The most intense combustion and the highest temperature occur just above the tuyere level; and, as the gases rise from this zone, a great deal of their heat is absorbed by the charge. The more fusible materials melt pretty well up in the column and, trickling over the refractory portions, flux the latter. It is this action, together with the effective contact of the hot gases with the charge, that accounts for the greater speed with which coarse material smelts in a blast furnace than in a reverberatory.

The burning of part of the pyrite, or sulphide of iron, yields iron oxide which, as in the reverberatory furnaces, combines with the silica, alumina and lime to form slag. The copper sulphide melts together with that part of the iron sulphide that has not been burned, to form matte. The two products collect in a shallow pool in the bottom of the furnace and flow out through water-jacketed spouts into large settlers. These spouts are so designed that the surface of the pool is maintained a little above the holes through which the molten mass passes, thus forming a trap to prevent any of the air blast escaping at this point. In the settlers the matte separates to the bottom, and the slag passes through an over-flow spout into a stream of water, which granulates it and carries it to waste. At suitable intervals the matte is drawn off through tap holes and delivered to the converters.

One of the smallest furnaces treats five hundred tons of charge per twenty-four hours, containing 8½% copper, and turns out eighty tons of matte with a content of 40% to 50% copper, and two hundred and seventy tons of slag; 4% of the charge is blown out by the air blast, but most of this is caught in the chamber and flue system; 60% of the total sulphur in the charge is burned off, amounting to thirty-three tons; 10% of the charge is also eliminated in the form of the carbonic acid gas that is driven off from the limestone.

The large furnace has a capacity of twenty-five hundred tons of charge per twenty-four hours.

The new economies effected in reverberatory smelting have made it unnecessary to use the blast furnaces except when more ore is mined than can be treated in the reverberatories.

Converting

Of all operations about the Works the converting process is the most spectacular. Flames of varying color are here seen belching from the throats of specially designed tilting furnaces of a pot-like shape; and, when these are turned down to rapidly pour off immense masses of slag or copper, there is a suggestion of Othello's "gulfs of liquid fire."

A converter, ~~as used~~ at this plant, is a large cylinder topped with a truncated cone, the cone being open at its upper, small end to receive the charges and as an outlet for the gases. It is made with a shell of heavy steel plate, thickly lined with magnesia brick. The inside dimensions are sixteen feet diameter and fifteen feet deep. At the back and near the bottom a line of tuyere pipes admits compressed air. As a converter with its charge of matte weighs three hundred tons and has to be frequently tilted, the supports are carefully designed and very substantially built. These supports are two pairs of massive steel rollers carried on equally massive foundations. Power for tilting is supplied through an ~~hundred~~ horse-power electric motor for each converter.

The converting process is the last stage in the series of eliminations that finally leave metallic copper. Molten matte is brought from the reverberatory and blast furnaces in large steel ladles, and poured into the converters—sixty-five tons to a charge. Air, at sixteen pounds pressure per square inch, is blown through the liquid mass, burning the sulphur and the iron. The sulphur dioxide gas passes into the flue, and the iron oxide combines with the silica and alumina of the raw ore that is added to supply these slag-forming substances. The slag is poured off into ladles by tilting the converter; and after practically all of the iron has thus been disposed of, the copper sulphide remains with a content of 78% copper. Continuation of the blowing burns off all but a very little of the remaining sulphur, and the metallic copper settles to the bottom alloyed with the silver and gold. This is finally poured into ladles and transferred to the refining furnaces. The conversion of a sixty-five ton charge of matte

Main Stack and Cottrell Treaters—Anaconda Reduction Works

requires three hours to slag the iron, and another hour and three-quarters to finish. The operators judge the stage of the process by the size and color of the flame, which changes from yellow through orange and red to blue, and also by the way, small particles of matte, slag and copper, which are thrown up by the air blast, act when they strike the hood at the lower end of the flue into which the converter gas passes.

The magnesia lining is one of the important recent improvements in copper metallurgy. In former practice the ore, supplied to flux the iron of the matte, was ground together with some of the clay-like slime that is produced in the mechanical concentration process, and tamped into the shell for lining. Such lining was rapidly consumed and had to be frequently renewed. This resulted in much greater expense than with magnesia lining, which is not appreciably subject to fluxing though it does very gradually suffer by mechanical wear from the wash of the fluid charge.

Unlike the slag from the reverberatories and blast furnaces, that produced in the converting process is too rich in copper to be thrown away. Most of it, still liquid, is taken directly to a special reverberatory furnace in the converter building. Suitable amounts of calcine, dried concentrate and fine siliceous material are put into this furnace, to produce a final slag that retains so little copper that it can be sent to the dump. That portion of the converter slag that is sent back to the blast furnaces is cast into slabs in a series of steel molds carried by a long chain conveyor, and allowed to solidify before being delivered to the stock bins.

To take care of the frequent handling of large ladles and of the "boats" that are used to charge the ore, there are three very large and powerful overhead traveling cranes, each equipped with electric motors, the operator riding in a cab that is hung on the crane.

Dust Chambers, Flues and Stack

All furnace departments are subject to some loss in the form of dust carried out by the draft. This dust amounts to only a very small percentage of the material treated; yet it carries considerable metal value. From most of the departments the dust-bearing gases pass directly into brick chambers so large that the coarser particles settle. These chambers are elevated above ground level, and the bottom is made up of a large number of steel hoppers from which the dust is drawn into cars and delivered to the reverberatory furnaces.

From the chambers the hot furnace gases pass through brick flues, most of which are twenty feet wide and fifteen feet deep. Altogether there is nearly a mile of flues of this size; and they all discharge into a main flue through which the smoke travels another half mile up the hill to the stack. For half its length this main carrier is sixty feet wide with walls rising twenty feet above the ground and with a V-shaped bottom extending seventeen feet into the ground. Still more dust settles in this large space, and is drawn through bottom gates into cars that run in a tunnel. The last half of this flue is one hundred and twenty feet wide and has a W-shaped bottom below ground, with two tunnels for discharging the excessively fine dust that settles here. This dust also is delivered to the reverberatory furnaces.

The flue gases finally discharge into the atmosphere through a stack which is built of perforated brick laid in acid-resisting mortar, and has numerous built-in bands of heavy reinforcing steel. It rests on a massive concrete foundation, towers five hundred and eighty-five feet above the ground and has an inside diameter of seventy-five feet at the bottom and sixty feet at the top, with walls varying from six feet thick at the base to two feet at the top. It has only recently been erected, being made necessary by the continued increase in capacity of the Works during the past few years, and has taken the place of an older one that was thirty feet inside diameter and three hundred feet high and which was the largest among large chimneys when it was built. The new one, which in its turn is now the premier among such structures, has six times the carrying capacity of the old and can easily discharge three or four million cubic feet of gas per minute.

Casting Anodes—Anaconda Reduction Works

Refining Into Anodes

Although the converters turn out the metallic copper, this product still contains small amounts of iron and sulphur, and also carries with it a little of the slag. These are removed in the refining furnaces, in which there is time and opportunity for more careful treatment than can be given in the converters. The refining furnaces are similar to reverberatory smelting furnaces but of smaller size; and like the latter, are heated with pulverized coal.

When one of these furnaces has been charged with liquid copper from the converter, the end of an iron pipe, carrying air under a pressure of sixteen pounds per square inch, is depressed below the surface of the bath. This air burns out the remaining iron and sulphur, the former entering the slag and the latter passing away with the furnace gases. From a two hundred ton charge of metal three tons of slag is scraped out of the furnace. This is rich in copper and is returned to the converters.

During the process of air refining more or less of the copper is oxidized, and considerable of this oxide dissolves in the bath. It is brought back to the metallic state by the reducing action of wood, the ends of green poles being forcibly held beneath the surface of the charge. The refiner determines when the copper has come to the right "pitch" by observing the form of the surface of small cooled samples which he takes in a little iron cup.

When the refining is completed, the copper is tapped from the furnace into cast-iron molds which are carried on a chain conveyor and which form the metal into anodes. These are cooled in a tank of running water, loaded directly into freight cars and sent to the Company's electrolytic plants where the silver and gold are separated.

THE ELECTROLYTIC REFINING OF COPPER

RARITAN COPPER WORKS

THE copper produced from the converter contains silver and gold from the original ore as well as many of the metallic impurities which are not completely removed during the smelting and converting operations.

In order to recover the silver and gold values, as well as to produce the exceedingly pure grade of copper required for electrical purposes, the copper, after melting in the shape of anodes, is shipped to one of the two refineries of the company for final refining. At these refineries the crude copper is refined by electro-chemical methods, and the refined product after melting is known as electrolytic copper to distinguish from copper that has not been so refined.

The Great Falls refinery is advantageously located, having an abundant supply of cheap electric power, essential for the economic operation of the electrolytic process. The Raritan Copper Works at Perth Amboy, N. J., is situated at tide water level to facilitate trans-Atlantic shipments, and is able to readily furnish copper to the large brass and rolling industries of the eastern part of the country. This refinery does a large custom refining of blister copper from various smelteries in Arizona, Utah, Mexico, Canada, Japan, South America, etc. The Raritan Copper Works also receives the silver and gold slimes from the Great Falls Refinery and in 1919, produced 19,503,573 ounces of refined silver and 120,629 ounces of refined gold, beside recovering as by-products other metals which originally were present as impurities in the crude copper.

The Raritan Copper Works

The plant of the Raritan Copper Works is located twenty-three miles from New York in the City of Perth Amboy, New Jersey, at the mouth of the Raritan River.

In addition to its lighterage service to New York and the foreign steamship lines, the Works has direct rail connection with the Pennsylvania, Central of New Jersey and Lehigh Valley railroads. Shipping facilities for both incoming and outgoing material are thus unexcelled.

The plant covers an area of forty acres and has a refining capacity of 40,000,000 pounds of copper per month. The materials received from the smelters are all in the form of crude bullion, no ores being handled. Most of the copper received comes from the Western smelters. Most of the copper products of the Anaconda Copper Mining Company of Montana and the entire copper products of the International Smelting Company of Tooele, Utah, the Inspiration Cons. Copper Company of Arizona, and the Miami Copper Company of Arizona, the Green Cananea Copper Company of Cananea, Mexico, the Pennsylvania Salt Company of Natrona, Penn., are refined at this plant. In addition to these sources, copper is regularly received from Rio Tinto, British Columbia, Copper Queen, Arizona Copper Company, Japanese and others.

The regular products of the plant include all commercial forms of refined copper such as wire bars, ingots, ingot bars, square and round cakes, wedge bars, slabs and billets. The by-products produced are refined silver and gold, platinum, palladium, selenium, tellurium, copper sulphate, and nickel salts.

Two grades of copper are produced. The highest grade or "N. E. C." Brand is the purest commercial copper on the market. It has a very high electrical conductivity and meets all requirements where copper of exceptional purity is desired. The second grade, "M. A." and "A. B. S." Brands is a pure ingot copper, and, while of too low conductivity for electrical uses, is an exceptionally satisfactory grade for casting purposes.

Silver, gold, platinum, palladium, selenium and tellurium are obtained from the anode

Bay of Tank House No. 2—Raritan Copper Works

slimes resulting from the electrolytic treatment of the copper anodes. Silver and gold are ordinarily disposed of in the usual bar form; platinum and palladium as sponge. Selenium is produced in several different forms, viz., red amorphous form, as powder; vitreous form as powder, sticks and half-pound cakes. Tellurium is made up into sticks, one-pound and five-pound cakes.

A small proportion of nickel is found in most crude copper. In the process of electrolytic refining this accumulates in the electrolyte and is separated out and crystallized at this plant, as pure nickel sulphate ($NiSO_4.7H_2O$).

The Furnace Department, in which crude copper and electrolytic cathodes are melted and refined, contains six refining furnaces with an aggregate capacity of about 2,300,000 pounds of copper per day and five anode furnaces with a total capacity of about 2,000,000 pounds per day. Waste heat boilers used on the furnaces generate a considerable proportion of the steam used in the plant.

Electrolytic refining of the anodes, received from the smelters and produced in the anode furnaces, is carried on in two tank houses. No. 1 Tank House contains 1,800 lead-lined tanks and is capable of turning out 21,000,000 pounds of cathodes per month. No. 2 Tank House, containing 1,656 tanks, will produce 19,000,000 pounds per month.

The electrolytic slimes resulting from electrolysis of the copper are worked up for silver, gold and by-products by fire and electrolytic methods. Silver and gold are parted electrolytically by the Thum and Moebius Systems, the gold being refined by the Wohlwill Process. Facilities are also provided for recoveries of the other by-products noted above.

Power for refining and general uses is supplied from two power houses. The aggregate capacity of the steam and electric generating units is above 14,000 K. W.

Shipping Refined Copper; View of Docks—Raritan Copper Works

General View of Plant—Raritan Copper Works

The Electrolytic Refining of Copper

The electrolytic refining of copper may be divided into three stages:

1. Melting the "Blister Copper" and casting into "Anodes" ready for the electrolytic tanks.

2. "Electrolytic" refining and obtaining "Cathode Copper" of 99.98% purity.

3. Melting of "Cathodes" and casting the commercial shapes, such as "Wire Bars" for wire mills, "Cakes" for rolling mills, "Ingots" for the brass and alloy manufacturers, etc.

The "Blister Copper" is delivered to the refinery in the form of slabs, weighing about 300 pounds. These are melted in reverberatory furnaces, known as "Blister Furnaces," of 500,000 pounds capacity. The details of the operation being practically the same as in the "Refining Furnaces" which will be described later. In this melting process but little refining is attempted although some of the impurities are removed with the slag and the sulphur is practically eliminated. The copper is cast into "Anodes" 37 x 28 x 1⅝ inches, weighing about 500 pounds. A typical assay of "Anodes" follows; a more complete analysis would disclose traces of many other elements.

Copper—	99.25%	
Silver—	70.0	ounces to the ton
Gold—	.25	ounces to the ton
Arsenic—	.060%	
Antimony—	.052%	
Nickel—	.050%	
Lead—	.053%	
Iron—	.058%	
Bismuth—	.003%	
Sulphur—	.004%	
Selenium—	.008%	
Tellurium—	.038%	
Oxygen—	.100%	

Theory of Electrolytic Refining

The theory of electrolytic refining is simple. If in an electrolyte of suitable composition is placed two copper sheets connected to the terminal of a generator, the sheet connected with the negative pole being known as the "Cathode," the other connected with the positive pole being known as the "Anode," on passing a direct current of suitable voltage and density, the current will enter the anode, pass through the solution to the cathode, dissolving copper from the anode and depositing an equivalent quantity of copper upon the

Raritan Copper Works

cathode. Owing to the energy generated at the anode, by the solution of the copper the total intensity factor, the voltage is small, usually about .25 volts per tank. The capacity factor, the amperage, follows strictly Faraday's law, copper acting as a divalent metal. One ampere per hour depositing 1.186 grams of copper.

In practice a number of tanks are connected in series with the electrodes in the individual tanks being in multiple. The electrolyte is an acid solution of copper sulphate usually 16% Bluestone with 12% of free sulphuric acid. The current dissolves the impure copper from the anodes, and deposits pure copper upon the cathodes, the impurities either dissolving and fouling the electrolyte, or not dissolving and dropping to the bottom of the tanks, known as "slimes." The products of electrolysis, therefore, are:

(*a*) Cathode Copper—99.98% purity.

(*b*) Anode Residues and Slimes.

(*c*) Impure Electrolyte.

Of the usual impurities found in Blister Copper, the nickel, cobalt, iron, zinc and arsenic dissolve and foul the electrolyte, while the silver and gold along with the other impurities remain in the slimes, and drop to the bottom of the tank.

Since the purity of the cathode copper depends on the purity of the electrolyte and the above impurities are constantly building up, it is necessary for all refineries to purify daily a certain volume of their electrolyte.

At the Raritan Copper Works the electrolyte is purified by a series of crystallizations, whereby the copper in the electrolyte is recovered as "Commercial Bluestone" and sold as such. The nickel, as a high-grade nickel sulphate (single salts) which finds a ready market and the acid after removal of the arsenic is used over again in the electrolyte.

The slimes are periodically pumped from the electrolytic tanks to the silver department, where, after an extended series of metallurgical processes, the silver and gold are recovered in a pure condition. The semi-rare metals, selenium and tellurium are also recovered, but the commercial demand for these two metals is so small that only a fraction of the possible recovery is attempted. Small quantities of valuable platinum and palladium concentrate in the slimes, which are recovered. The lead, bismuth and antimony are shipped as crude metal to lead smelters for further purification.

Lifting Cathodes out of Tank—Raritan Copper Works

Practice of Electrolytic Refining

At the Raritan Copper Works the multiple system of refining copper is employed. In this system the anodes and cathodes in the tanks are connected in multiple, the tanks themselves being connected in series.

The electrolytic tanks are of wood, 118 x 34 x 44 inches, and are lead lined. They are carefully insulated on glass and are so placed that any leakage of electrolyte may easily be detected. There are two tank houses, one of 1,800 tanks and the other of 1,656 tanks. Four circuits of 7,000 amperes in each tank house, which gives a current density of 18 amperes per square foot of cathode surface. These tanks are arranged in nests of 11 cells each, the nests being so placed together so as to circulate the electrolyte in cascade through two tanks. The circulation of the electrolyte being maintained at the rate of four gallons of solution per minute per tank. After leaving the lower tank in the cascade, the solution drops to "hot wells" where it is reheated by means of steam coils to 60° C. and from there pumped to the top of the cascade. Experiments having shown that the best conditions for the electrolysis is a circulating solution at 55° C.

Each tank contains 28 anodes and 29 cathodes. The anodes weighing 500 pounds apiece, properly spaced, are lowered together by means of a crane into the tanks. Between the anodes are then placed the cathodes, thin "Starting Sheets" of copper, 30 inches wide and 37 inches long, weighing about eight pounds. At the end of ten days with a current of 7,000 amperes the cathodes have increased in weight to an average of 135 pounds each. They are then taken out of the tanks, carefully washed from adhering electrolyte and sent to the "Refining Furnaces" for melting. A second set and later a third set of "Starting Sheets" follow the same procedure. At the third lifting of cathodes, 30 days from the time they were put into the tanks, the anodes of crude copper have so dissolved that only a skeleton of the original anodes remain—"Scrap Anodes."

The solution is syphoned from the tanks, the "Scrap Anodes" washed from adhering slimes and sent to the "Blister Furnaces" to be recast into 500 pound anodes. The slimes are then pumped to the Silver Department where the gold, silver and other valuable by-products are recovered.

The cathode starting sheets are made on blanks, "Starting Blanks" of hard rolled polished copper ⅛-inch thick. These blanks become cathodes in regular electrolytic tanks and in 24 hours enough copper has deposited on them so that it may easily be stripped off in the form of thin sheets of copper, which, after being properly looped, become "Starting Sheets." The copper deposits on both sides of the blank so that one "Starting Blank" makes two "Starting Sheets."

In order to prevent the deposited copper from adhering fast to the "Starting Blanks" the blanks are covered with a thin film of grease or oil before electrolysis. The blanks are grooved ½-inch from the edges; the deposited copper parting readily along the groove, giving a standard size "Starting Sheets" without trimming. After stripping the copper remaining on the edges of the blanks, the blanks are oiled and again put in the electrolytic tanks, the "Starting Blanks" being in continuous use.

The electrolyte is an acidulated solution of copper sulphate. Analyzing—4% copper with 12% of free sulphuric acid. The sum total of the impurities such as nickel and arsenic, etc., is kept below 2%.

The operation of the tank house requires constant vigilance. The Power House delivers to the tank house a given quantity of current, for which the tank house is supposed to furnish an equivalent weight of cathode copper. Theoretical figures all being based on Faraday's law, but short circuits, bad contacts, leakage of current all tend to lower the efficiency and are carefully watched. Most refineries obtain an efficiency of 90% based on power consumption.

Melting Copper Cathodes

The cathode copper, containing as it does 99.98% copper, should be an easy melting proposition, but copper when molten absorbs reducible gases, more particularly sulphur, to such an extent that the melting process becomes a refining process, refining from the gases absorbed during the period of melting. The presence of oxygen, or rather copper sub-oxide, greatly increases the solubility of these gases in cast copper and it is for this reason that from .03% to .08% of oxygen will always be found in commercial copper.

The copper is melted in large reverberatory furnaces of 225–250 tons capacity. The furnaces are lined with magnesite bricks up to above the metal line, with silica brick roof and sides. The fuel is bituminous coal or oil and the gases on leaving the furnaces are conducted through waste heat boilers. The melting or refining operation may be divided as follows:

(1) Charging.
(2) Melting.
(3) Oxidizing.
(4) Poling.
(5) Casting.

All operations are so conducted that it takes about twenty-four hours to work the complete cycle of a charge.

(1) Charging. Charging was formerly done by hand, a laborious operation, but now all furnaces are charged with a charging machine. The machine picks up 40 cathode sheets weighing about 5,000 pounds at a time and puts them in the furnace much more compactly than could be done by hand, thus materially increasing the capacity of the furnace. The time of charging a furnace is usually two hours.

(2) Melting. After the furnace is charged, the side doors are closed, luted, and fire increased to hasten the fusion of the copper. In about 10 to 12 hours enough copper has melted so that the oxidation period can begin.

Casting Side of Refining Furnace—Raritan Copper Works

(3) OXIDATION PERIOD. In this period the molten copper is nearly saturated with cuprous oxide and brought to what is known as "Set Copper" containing from 4% to 5% of suboxide of copper. In this process all impurities are oxidized and all hydrogen, carbon monoxide and sulphur compounds removed.

This is accomplished by blowing compressed air through iron pipes under the molten bath, also by "Flapping" the charge. Flapping consists in striking the surface of the metal with the edge of the head of a rabble, in order to uncover the surface of the copper.

The furnace man watches the operation by taking button samples and observing the fracture. The fracture in all stages being characteristic and well defined.

Charging Side of Refining Furnace Raritan Copper Works

(4) POLING. After the charge is oxidized, the slag formed is skimmed off. The bath covered with coke or charcoal and the "Poling" process begins. This consists of forcing the butt ends of green poles underneath the surface of the metal. This causes a violent action in the metal bath, and the carbon from the wood and coke quickly reduces the dissolved cuprous oxide. This operation takes two to three hours. When the suboxide of copper has been reduced to three-tenths to six-tenths per cent, which can be readily told by the fracture of the button sample, the copper has reached the "Tough Pitch" stage and is ready for casting.

(5) CASTING. This operation consists of casting the molten copper into commercial shapes, wire bars, slabs, wedge bars, cakes,

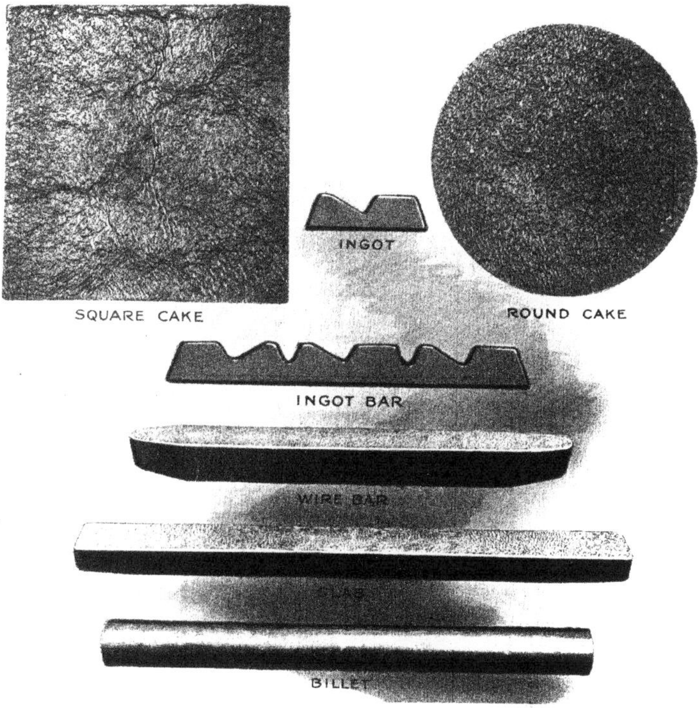

Shapes of Refined Copper

ingots and ingot bars. The molds are made of copper and sprayed with a bone-ash emulsion. The molds are placed radially on a circular casting machine, "Clark Casting Machine." As the wheel turns around, the molds are filled with copper from the furnace. After the copper has set, the molds automatically tip, dropping the copper bars into a water bosh, which pickles the copper and gives it the characteristic red color. The bars come up from the bosh on an endless belt, and after weighing, are ready for shipment. A representative anyalsis of refined electrolytic copper would be somewhat as follows:

Copper—	99.94%–99.97%
Oxygen—	.02%–.05%
Silver—	.0010%
Gold—	.00001%
Sulphur—	.0030%
Iron—	.0025%
Nickel—	.0015%
Arsenic—	.0015%
Antimony—	.0020%

The sum of the metallic impurities being less than one-hundredth of 1%.

Through all furnace operations considerable slag is made, mainly an oxide slag containing 40% to 50% of copper. This is smelted in a Blast Furnace to Blister Copper.

The Commercial Shapes of Refined Copper

The cathode copper, after melting and refining in the refining furnaces, is cast into the commercial bars best suited as to weight and dimensions for the purpose it is intended to be used for. Refined copper is the basis of a great many industries such as the manufacture of brass and other alloys, copper rolling mills, wire drawing mills, tube mills, etc., and each industry has designed and specified for its own particular use, special shapes of refined copper, best suited for its own purpose.

Ingots of copper are used primarily where the copper has to be remelted in crucibles either for the making of copper castings or the manufacture of alloys such as brass, bronze, nickel silver, etc. Ingots, therefore, are of a shape that will readily fit into crucibles, and are about ten inches long, weighing from sixteen to twenty-two pounds. Ingots have either one or two notches, to facilitate breaking the ingots in two or three pieces, if necessary. Ingots being so small are difficult to handle and ship in large quantities, and ingot bars are often used instead. An ingot bar may be considered as three ingots in one bar, which, owing to the deep notches, are readily broken apart. Ingot bars are about thirty-two inches long and weigh from seventy to eighty pounds.

Wire bars, the most popular form of refined copper bars, are used by wire mills for the drawing of copper wire; these bars are cast with pointed ends in order to facilitate the bar taking the first set of rolls. The size and weight of wire bars vary greatly, the length from 38 to 100 inches and the weight from 135 to 770 pounds. The 200-pound bar, 54 inches long, is the size commonly employed.

Slabs and square cakes of various sizes are used for rolling purposes, where sheet copper is the final product. The size of these bars vary greatly, depending on the size of the finished product.

Circular cakes are used for the manufacture of large seamless, cylindrical products such as hot water heaters, tanks, etc.

Billets are used for the manufacture of seamless copper tubing of all sizes, and during the war were used extensively in the manufacture of the smaller sizes of shell bands. Billets vary from 2½ to 8 inches in diameter, 15 to 50 inches in length and from 70 to 600 pounds in weight.

Copper Storage—Raritan Copper Works

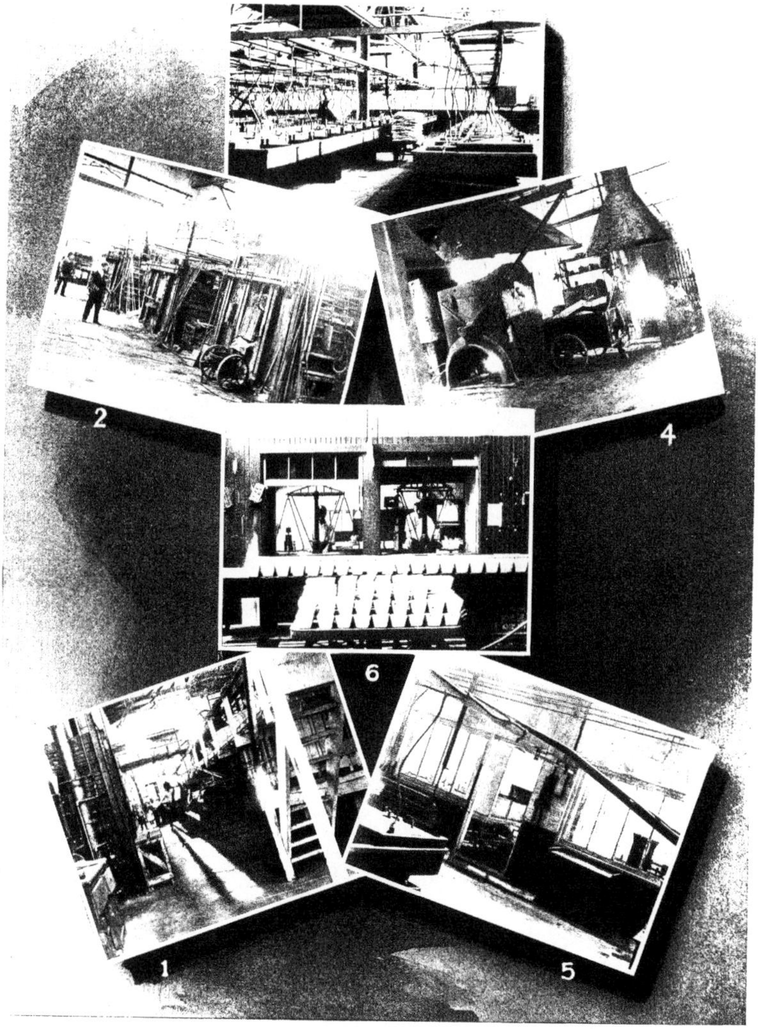

Views of Silver Department—Raritan Copper Works

No. 1—Slime Department. No. 2—Doré Furnaces. No. 3—Parting Department, Thum Sections. No. 4—Silver and Gold Melting Furnaces. No. 5—Gold Refining, Wohlwill Process. No. 6—Weighing and Shipping Department.

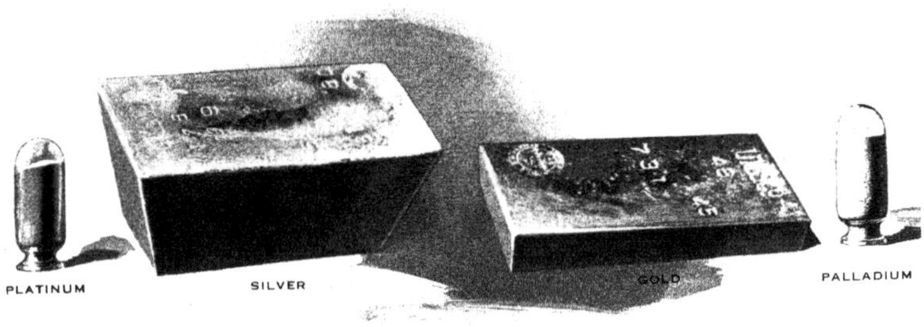

Precious Metals—Raritan Copper Works

Precious Metals

The functions of a copper refinery are twofold, first to produce the high grade of copper metal, which the electrical industry is dependent upon, and second to recover from the blister copper the precious metals which it contains. It is estimated that about 80% of the total silver, and 15% of the total gold production of the world are recovered as by-products from copper, lead, nickel and cobalt refining.

The Raritan Copper Works not only recovers the precious metals from the blister copper it refines, but also receives for treatment the slimes from the copper electrolytic refinery of the Great Falls Reduction Department, and the doré bars (crude silver containing gold) from the lead refinery of the International Lead Refining Company, thus becoming the precious metal center of the Anaconda Copper Mining Company. For the year 1919, this refinery treated for all the plants of the company 165,442 tons of copper bullion, and 6,514,987 ounces of silver from which there were produced 326,671,962 pounds of refined copper, 19,503,573 ounces of fine silver and 120,629 ounces of pure gold.

In the electrolytic refining of copper, all the precious metals originally in the anodes drop to the bottom of the tanks during the process of electrolysis and are known as "Slimes" or "Anode Mud." When the scrap anodes are lifted from the tanks after thirty days of electrolysis, the slimes are sluiced from the bottom of the tank and pumped to the silver department. The slimes will vary considerably in composition, depending upon the grade of blister copper refined. From Anaconda anodes, the slimes will have about the following composition:

Silver—43.23%—12,610 ounces per ton
Gold— .234%—68.4 ounces per ton
Copper—13.86%
Arsenic— 3.88%
Antimony— 2.46%
Selenium— 1.46%
Tellurium— 6.14%
Lead— 3.96%
Bismuth— 0.26%
Iron— 0.22%
Nickel— 0.27%

The slimes on reaching the silver department are allowed to settle in large settling tanks, the liquor being sent back to the tank house, and the slimes filter pressed to a solid cake containing about 25% of water. In order to remove the copper, the slimes are given a light roast (300° C) in an oil fired oven. This converts all the copper into copper oxide, which is readily leached with 10% sulphuric acid in "Agitators," lead-lined iron tanks with mechanical paddles to keep slimes and solution agitated. The slimes, after leaching, contain less than 1% of copper, and after again filter pressing, are ready for the fire treatment.

This treatment consists of melting and refining the slimes in small oil fired reverberatory furnaces, so called "doré furnaces."

During the refining process the lead forms a lead slag which is skimmed off and sent to the Lead Refinery for further treatment.

Selenium and tellurium are partly

volatilized, but are mainly removed by the addition of alkaline fluxes, soda ash and nitre. Arsenic and antimony are volatilized and recovered from the flue dust. The furnace treatment will take about 30 hours for a charge of 10,000 pounds of wet slimes, and will result in about 3,000 pounds of doré metal containing 98.5% of silver, 0.5% of gold, with a small amount of metallic impurities. The doré silver is cast into anodes, and electrolyzed, using either the Thum or Moebius system.

In the Thum system the electrolytic cells consist of shallow, glazed, porcelain tanks, the bottom of which, lined with graphite slabs, forms the cathode. The doré silver anodes are arranged horizontally in wooden frames above the cathode. These frames are covered with canvas, which acts as a diaphragm and prevents the slimes from mixing with the cathode deposit. The silver is deposited in a loose crystalline form which is scraped off from the bottom of the tank at periodical intervals. The electrolyte for both the Thum and Moebius system is a neutral solution of silver nitrate containing about 4% of silver; any copper in the silver anodes dissolves and gradually builds up in the electrolyte. In the Moebius system the anodes and cathodes are arranged vertically in much the same manner as in copper refining, the anode being inclosed in canvas bags in order to catch the gold slime, the cathodes are of thin, rolled silver and mechanically operated wooden scrapers are used to knock off the silver crystals from the cathodes into trays at the bottom of the tanks which are periodically withdrawn and emptied. The cathode deposit of silver crystals is very pure, and after washing is melted in large graphite retorts into standard thousand ounce bars, assaying over 999.2 in fineness.

The slimes resulting from the silver electrolysis contain all the gold, platinum and palladium, which originally were present in the blister copper, and after treatment with sulphuric acid to remove the excess of silver present, are melted in graphite crucibles and cast into gold anodes, and electrolyzed using the Wohlwill process.

This process which is used for the refining of gold in all the United States mints, consists in electrolyzing gold anodes, in a hot acid solution of gold chloride. The solution, which is kept at 70° C., contains from 7% to 8% of gold, and 10% of hydrochloric acid, the density being 1.30. In order to keep down the interest charges on the gold tie-up, a high current density, 110 amperes per square foot of cathode surface is used; this necessitates a thorough stirring of the electrolyte, which is accomplished by means of a small air lift. The cathodes are made up of thin ribbon of rolled gold, and the deposit which adheres firmly to the cathodes is exceedingly pure, averaging 999.75 gold fineness. This deposit is melted in graphite crucibles and cast into standard bars.

In copper and silver electrolysis the more noble metals have been insoluble and have formed the slimes. In the gold electrolysis, however, the platinum and palladium are totally soluble and are allowed to accumulate in the electrolyte until present in sufficient quantity to recover. Chemical methods, following the ordinary quantitative analytical separation of these metals are used in the process of recovery, depending upon the precipitation of both metals when in an oxidized condition by ammonium chloride to separate them from the bulk of the gold in the electrolyte.

The amount of platinum and palladium recovered is small, averaging only about one ounce of the combined metals per million pounds of copper refined, yet owing to the large tonnage of copper refined, the high prices of these metals and the simplicity of the process, the recovery is commercially successful.

Other metals recovered in the process of copper refining are nickel, selenium and tellurium. The lead, antimony and bismuth are shipped as crude metals and slag to the lead refinery for further treatment. Owing to the large tonnage of copper refined, it is imperative that a refinery removes from the system, the impurities found in the blister copper as fast as they come in.

On a monthly production of 40,000,000 pounds of refined copper, one-hundredth of 1% of any impurity means at the end of a year 48,000 pounds of that particular impurity, which, if not removed from the system, is liable to form a serious clog in the economic operation of the process.

THE ROLLING OF COPPER INTO WIRE

GREAT FALLS REDUCTION DEPARTMENT

Historical

EARLY in 1891 the Boston & Montana Consolidated Copper & Silver Mining Company broke ground for a copper reduction works on the north bank of the Missouri River, across the river from the east end of the City of Great Falls, and about 2.5 miles from the business district of that city. Black Eagle Falls are at this point in the Missouri River. These works were built to treat ore from the Company's mines in Butte yielding copper and relatively small amounts of silver and gold.

About a year later, or in March, 1892, a concentrator was in condition to begin operations, and shortly after this roasting furnaces and reverberatory smelting furnaces were finished. Bessemer converters were also installed about this time. This completed a chain of operations for the production of blister copper, excepting that it was found necessary to erect a blast furnace plant for the retreatment of converter slag. The blast furnaces were completed and in operation in 1893.

In 1893, an electrolytic copper refinery and a furnace refinery were completed and started in operation, making it possible at that time to carry the treatment from ore to finished commercial shapes of refined copper.

During the year 1910, the properties of the Boston & Montana Copper & Silver Mining Company were taken over by the Anaconda Copper Mining Company, and since then the works at Great Falls have been known first as the Boston & Montana Reduction Department, and more recently as the Great Falls Reduction Department of the Anaconda Copper Mining Company.

In 1916, a plant was completed for the production of electrolytic zinc from ore mined in Butte and concentrated in Anaconda.

In 1918, a mill was finished for the manufacture of copper rods, wire and cable, which are made from the refined copper produced by the furnace refinery. The completion of this mill brings the copper industry in Great Falls to the point of a product ready for the consumer.

Also, in 1918, a ferro-manganese plant was completed for the production of ferro-manganese from Butte ore.

During a few years prior to 1918 the work of copper concentrating and smelting, for which the original works were built, was gradually reduced, until, in 1918, this work was discontinued entirely, so that now the operation is changed from copper concentrating and smelting to that of copper refining, copper rod, wire, and cable manufacture, electrolytic zinc production, and ferro-manganese production. The copper smelting equipment remains in place and may be used when there is occasion for doing so.

Electrolytic Copper Refinery

The first electrolytic copper refinery, which was built in 1892, continued in operation until 1916, when a new modern well-equipped refinery was built.

The new refinery, which began operating in March, 1916, consists of the main tank room 382 feet 8 inches long by 251 feet wide, a sub-station adjoins the tank house, as well as a sulphate building for the purification of the electrolyte, and the preparation of the electrolytic slimes for shipment.

The tank house, which is operated on the multiple system, contains 1,020 refining tanks. The tank room is divided into four crane bays of 240 tanks each, sixty tanks being devoted to the production of starting sheets. Each tank contains twenty-five anodes and twenty-six cathodes. The anodes weighing 575 pounds each remain in the tanks twenty-four days during which time four drawings of six day cathodes are made.

The tank house is divided into two electrical circuits of 510 tanks each. Two motor-generator sets on each circuit deliver 9,800 amperes at 210 volts, an equivalent of twenty-four amperes per square foot of cathode surface, and a production of 545,000 pounds of

General View of Works from Southeast—Great Falls Reduction Department

cathode per day at an ampere efficiency of 88% to 90%.

The furnace refinery is equipped with two reverberatory furnaces, each with a capacity of 400,000 pounds of copper per charge, each furnace being provided with charging machines, waste heat boilers, and an improved Clark casting wheel.

Rolling Mills

The building for the rolling mills was begun in January, 1918, and was completed in May, 1918, the foundation for the rolling mills and other machinery having been put in before the erection of the building was begun. It is 450 feet long by 125 feet wide, and is constructed of steel, brick and glass. Ample provision for light and ventilation has been made. This building contains all the various sections of the plant, as follows:

The rod mill, consisting of an 18-inch roughing mill, intermediate mill, finishing mill, bar-heating furnace, coilers, roll lathe for grooving the rolls, and all other accessory apparatus necessary to make the mill complete in operation and handling the product.

The wire-drawing machinery, consisting of a 4-block bench for drawing large sizes of wire in coils, a continuous trolley wire-drawing bench with its reel stand, one 10-die continuous wire-drawing machine for drawing No. 12 and smaller sizes of wire, in coils or on reels, one 6-die and one 5-die machine for drawing wire from No. 6 to No. 12, and three single-die machines for drawing medium hard wire.

The annealing equipment, consisting of one double chamber non-oxidizing annealing furnace and one open-fire furnace.

The stranding equipment, consisting of three stranding machines of different capacity.

All of these groups of machines are served by a 5-ton traveling crane, which covers the full width of the building and runs its entire length.

Electrolytic Copper Refinery—Great Falls Reduction Department

Wire Mill—Great Falls Reduction Department

Rod Mill

The wire bars from the furnace refinery are first placed on a table in the rear of the bar-heating furnace, and a pusher, operated by compressed air, moves them along into the furnace, which holds 100 bars lying side by side throughout its length. It is heated by fuel oil burners located at the opposite end from that at which the bars enter.

The bars are taken out of the furnace through a door, located near the heating chamber. As fast as they are taken out more bars are pushed in at the rear end and the bars already in the furnace are moved toward the heat and the discharge door.

The heat travels the length of the furnace. The smoke and gases are taken out through a flue at the rear end which connects with a stack located outside of the building.

As the bars are taken out of the furnace at the rate of 100 an hour, it takes an hour for a given bar to travel through the furnace. In this way the heating takes place gradually and can be controlled so as to have each bar at the proper rolling temperature when it reaches the discharge door.

The bars are taken out at the discharge door with a pair of tongs suspended from a trolley which runs in line with the first groove in the rolling mill. This first groove is in what is known as the roughing mill, which consists of three rolls eighteen inches in diameter and sixty-four inches long, one above the other, driven from a motor through a reducing gear unit and a set of pinions. The direction of rotation of these rolls is such that the bar, after entering the first groove, which is in the top and middle rolls, passes through this groove and drops down into position for entering the second groove, which is in the middle and bottom roll, and passes back through this second groove to the side of the mill from which it started. It is then raised into position for entering the third groove, which is next to the first groove. In this way it passes back and forth through the mill seven times, each pass reducing the cross section of the bar and increasing its length.

After leaving the roughing mill the bar passes to the intermediate and finishing mills, consisting of five and six pairs of rolls respectively, each alternate pair rotating in opposite directions. When the rod, as it comes from the roughing mill, passes through the first pair of rolls in the intermediate mill, a man catches the end of it with a pair of tongs and passes it back through the next pair of rolls, the rod running in a loop on the inclined iron floor, which is on both sides of the rolls. This process is repeated until the rod has run through all the different pairs of rolls. As it leaves the final pair, which gives it its finishing size, it passes through an iron pipe into a coiling machine, where it is wound into a coil.

The finished coils are black with the oxidation caused by the exposure of the hot rod to the air during the rolling process. This oxide is removed by placing the coils in a dilute solution of sulphuric acid. After washing in water, the coils are then ready to be drawn into wire.

As rolled, the rods vary in size from one-fourth to five-eighths of an inch in diameter, the larger rods being used for drawing into larger sizes of wire and the smaller rods for smaller wire.

Wire Drawing Machinery

The wire drawing process consists of drawing the rod through a succession of dies until its diameter has been reduced to the diameter of the wire required.

For the larger sizes of wire the rod is drawn through one die at a time until it is finished, but for smaller wires the rod is placed on a continuous wire-drawing machine and is drawn through a succession of dies at the same time.

These machines have a series of drawing rolls, each of which draws the wire through one die, after which it passes through the next smaller die and on to the next drawing roll, this being repeated until it passes through the finishing die. It is then either drawn into a coil on a revolving block, or drawn or wound on a reel which is so driven as to take the wire as fast as it is drawn through the last die. The drawing rolls and blocks on these machines run at increasing speeds proportioned so as to take care of the increasing length of wire produced by the elongation of drawing.

General View of Interior of Wire Mill looking Southwest Great Falls Reduction Department

Annealing

The process of drawing the rod through the dies to the finished size hardens the copper about in proportion to the reduction in the diameter. Wire drawn on these machines is shipped as hard drawn wire or is passed through an annealing furnace which anneals it and renders it soft and pliable, which is known as soft drawn wire.

Medium hard wire is produced by drawing the rod to a certain size which, after being annealed, will require just the necessary amount of further drawing to produce the degree of hardness specified.

The dies used for wire drawing are small circular dies made of chilled cast-iron, cast with a tapered hole, and are reamed to exact size by hand. After the hole in the die becomes too large for certain sizes of wire it is then reamed out to a larger size and used for larger wires, this process being repeated many times.

The usual variation in diameter allowed on all wire of sizes No. 10 and smaller is one one-thousandth of an inch. Accuracy as to size is required and it is necessary to shape the die so that it will hold its size within this limit after withstanding the wear of drawing a wire as long as four miles from one rod.

For making trolley wire to be furnished in long lengths, it is necessary to join a number of rods, which is done by brazing them with silver solder before drawing. A sufficient number of rods are brazed so as to produce a certain length of finished wire. In many cases this length is one mile, and the weight of the wire varies from 1,687 pounds for 1/0 size to 3,382 pounds for 4/0 size. After being brazed the rods are drawn through two or more dies continuously and are wound on a reel at the same time.

The dies used for making round trolley wire are made of chilled iron, the same as the dies for making smaller wires.

For grooved, figure 8, and other shapes of trolley wire, the dies are made of the best quality of steel suitable for this purpose. They are carefully punched and sized, then hardened and polished. Under the best conditions a die will draw about five miles of wire, after which it has to be remade.

Stranding Equipment

In the manufacture of strand or cables, the wire composing the strand or cable may be hard, medium hard or soft.

The wire is either drawn on iron reels or wound on reels from coils and these reels are placed in the stranding machines, which are of different types and the capacity of which is rated as the number of wires that the machine will lay up into a cable.

Strand such as is used for power transmission lines is made on a high-speed machine which will lay up six wires around a center wire. This is called a 7-stranding machine.

Bar Heating Furnace at Wire Mill—Great Falls Reduction Department

Stranding Machine at Wire Mill—Great Falls Reduction Department

Another machine consists of two revolving circular frames which usually revolve in opposite directions and in which the iron reels containing the wire are placed. The first frame holds six reels and as this frame revolves these six wires are laid around a center wire which passes through the center of the frame. These seven wires then form the core of the cable and pass through the center of the next frame. The twelve wires which this second frame holds are laid around this core of seven wires, composing a 19-wire strand or cable.

Another machine of the same type, but with three frames holding 6, 12 and 18 reels each, makes a cable of three layers or a total of 37 wires.

If required, the cable of 37 wires is passed through the center of this machine and another layer of 24 wires is added, making a 61 strand cable.

It is also possible to pass this through, adding 30 wires if a cable of 91 wires is desired.

The complete cable passes through a revolving drum which takes it up as fast as it is twisted. It then passes from the drum to the reel on which it is to be shipped, which is driven so as to take it from the drum at the proper speed.

The revolving frames are driven through reversing and interchangeable gears, as is the take-up drum. The relation of the speed of the frame to the speed of the drum determines the pitch or lay of the wires in each layer, this lay usually varying with the number of wires.

It is interesting to note that the products of the rolling mills department are the first of their kind to be manufactured west of the Mississippi River. Previous to the erection of this plant, the wire bars from the furnace refinery were all shipped to mills in the East and Middle West. When wire and strand were required for the numerous power-development projects in this and other far western states it had to be shipped back here from the East. In many cases this wire and strand were made from this Company's wire bars. Hence, as the wire bars are made from copper produced by this Company's own mines, and as the rods, wire, cable, etc., are manufactured at its own plant, it is now permissible to say "From mine to finished product."

General View of Electrolytic Zinc Plant from the South — Great Falls Reduction Department

ELECTROLYTIC ZINC

GREAT FALLS REDUCTION DEPARTMENT

Electrolytic Zinc Plant

This branch of the Great Falls Reduction Department is divided into four main divisions with their subsidiary departments, roasting, leaching, electrolyzing, and casting. The rated capacity of the plant is 150 tons of grade A zinc per twenty-four hours.

Roasting Division

The zinc roasting furnaces are located in the general smelter building, on the north side between the blast furnaces and the reverberatories.

There are fourteen roasting furnaces of the Anaconda type, a modified Wedge-McDougall furnace, in two rows of seven each. Each furnace consists of a brick-lined steel shell, twenty-five feet in diameter, has seven hearths and one open-dryer hearth on top, a 5-foot, brick lined, hollow, revolving center shaft which carries twenty-six water-cooled arms, four for the dryer and first hearths and three for each of the other six hearths.

Each furnace has an individual motor, direct connected to a series of enclosed reducing gears from which the center shaft is turned by a bevel pinion. The water for cooling the arms is delivered through a water column revolving with the center shaft and connected to the water supply mains by a swivel joint.

The material to be roasted consists of zinc concentrates shipped from Butte and Anaconda in 50-ton, bottom-dump, steel cars and delivered to a battery of fourteen storage bins having a total capacity of approximately 3,750 tons, located just under the "high line" tracks. From these bins the material is loaded into 56-cubic-foot, V-shaped, rocker dump cars of 36-inch gauge and hauled in trains of eight cars by an 8-ton electric locomotive to individual feed hoppers immediately above the furnaces. Each furnace is equipped with an apron feeder driven by gearing to the revolving center shaft, and designed so as to regulate the rate of feed as desired. The rabbles on the arms (previously mentioned) are so set as to move the material from the circumference to the center of the furnace, and vice versa, on alternate hearths, delivering it, finally, into the calcine hoppers under the furnaces. From these hoppers the hot calcine is drawn into electrically operated, 4-ton, bottom discharge larry cars, hoisted by means of an 8-ton Otis elevator and discharged into revolving coolers, which in turn discharge into a 400-ton storage bin for shipment by the local electric tram to the zinc leaching plant.

The object of roasting the concentrates is to convert the zinc and lead sulphides into the sulphate and oxide forms, expel the excess of sulphur and to oxidize the iron and other metals in the ore. This process thus renders a product which is suitable for subsequent treatment in the zinc leaching plant.

Each furnace has two fire boxes, one on each side of the seventh hearth. The fuel used is pulverized Lochray coal, delivered to the roaster storage bin by means of screw conveyors from the coal pulverizing plant. From the storage bin the coal is delivered by screw conveyor to individual furnace hoppers, each of which is equipped with a feed screw driven by a variable speed mechanism which in turn is driven from a central main line shaft belt driven by an electric motor. The coal is blown into the fire boxes under 12-pound air pressure through especially designed burners. The ash is disposed of through grizzlies to launders and sluiced to the river.

The gases are carried in cross flues to a main building flue and thence to the up-take and cross-take to the main dust chamber, where the dust is settled to be subsequently drawn and reroasted. The gases pass from the dust chamber through a connecting flue to the stack.

The zinc leaching division requires about 600 tons of roasted concentrate per day.

Zinc Roasting Furnaces—Great Falls Reduction Department

Leaching Division

The roasted zinc concentrates arrive in standard gauge steel cars at the highest level in the building and are dumped into steel bins, each bin serving four leaching tanks. There is a small compartment in each bin for ground limerock to be used in the leaches.

The leaching solution which is the return solution from the electrolyzing division, or spent electrolyte, is elevated by hard lead, centrifugal pumps through lead pipe lines to storage tanks slightly above the level of the top of the leaching tanks. From these tanks the solution is distributed to the separate leaching tanks through lead lines, each leaching tank having a connection from one of these lines with a hard lead valve for regulating the flow.

The leaching tanks are of the Pachuca type operated in series. Acids solution is run into the first tank which discharges into the top of the second tank where some roasted concentrates are added. The second tank discharges into the top of the third tank and so on through the series, the final discharge being into Dorr classifiers for removal of coarse material. Roasted concentrates are added to the leaching tanks until all acid is neutralized and iron, arsenic, silica, alumina, etc., are precipitated. The overflow from the Dorr classifiers goes to Dorr thickeners for the separation of solids and solution.

The Dorr thickeners have redwood staves bound together by lead-covered iron hoops. The mechanisms are made of bronze and lead-covered steel, no iron or steel being exposed to the solution. The thickened pulp is drawn from the bottoms of these tanks and is again treated with acid in Pachuca tanks which discharge into additional Dorr thickeners. The thickened pulp from these thickeners goes to Oliver filters, the filtrate being returned with the thickener overflow to the first leach. The Oliver cake goes to the residue dryers.

The overflow from the first set of Dorr thickeners goes to the purification system for removal of copper and cadmium by agitation with zinc dust. The precipitated copper and cadmium and unused zinc dust are separated from the solution in settlers and plate and frame presses. After purification and clarification the zinc bearing solution is pumped to storage tanks which supply the electrolyzing division.

Electrolyzing Division

In this division, the zinc is deposited from its solution on aluminum sheets, from which it is stripped at regular intervals. The cells in which this deposition takes place are divided into units, the power for each unit being supplied by a rotary converter. The cells are lead lined and are arranged in a series of cascades. The solution feed system is so arranged that fresh solution may be added to any cell in the division.

The solution as fed to the cells contains no free acid, but, as the zinc is removed from solution, acid is formed and the solution leaving the cells contains approximately one and one-half times as much acid as there was zinc removed in the cells. This solution, low in zinc and high in acid content, becomes the leaching solution in the leaching division.

The deposited zinc is stripped from the aluminum sheets by hand and is loaded onto small cars that are pushed by hand to the casting division.

Casting Division

Here the zinc sheets from the electrolyzing division are melted and the molten zinc is poured into molds bearing the "Anaconda Electric" trade-mark. Each slab weighs about fifty pounds and is stamped with a lot number covering that day's production. Before shipment, each lot is certified to contain 99.9% zinc. In melting the zinc sheets there is a certain loss incurred in the formation of dross which is separately treated for the recovery of its zinc content.

All of the cathodes are melted in coal-fired furnaces. All casting is done by hand from ladles carried by trolleys on overhead beams, and the slabs are loaded by hand onto trucks that are pulled by hand to cars or storage docks.

Zinc Residue Dryers

The residue from the leaching division is conveyed from the Oliver filters to coal-fired horizontal, rotary dryers, the dried product being sent to the smelter department.

A flue system and stack are connected to

Electrolytic Zinc Plant; Oliver Filters—Great Falls Reduction Department

Electrolytic Zinc Plant; Drawing Zinc—Great Falls Reduction Department

Electrolytic Zinc Plant; Casting Zinc—Great Falls Reduction Department

handle the gases from the dryers and serve to keep the dust loss at a minimum.

Ferro-Manganese Plant

The ferro-manganese plant is located at the west end of the main smelter building. There are five electric furnaces, each with a capacity of twenty tons daily. The furnaces are operated on three-phase, 60-cycle current and each furnace is provided with a separate transformer rated at 5,630 kilo volt-amperes. The electrodes are of amorphous carbon, two electrodes in a common holder being used on each phase. The electrodes are twenty-four inches in diameter and are operated on the continuous-feeding principle. The furnaces are provided with automatic regulators of the Seede type for maintaining the lead constant.

The charge consists of manganese ore coal, or coke, iron ore, and limerock in proper proportions. The weighing and mixing of the charge is almost automatic. Sufficient charge may be mixed on day shift to run the plant during the entire twenty-four hour period. The mixed charge is delivered by conveyors into a mixed-charge storage-bin with a capacity of 500 tons, from whence it is loaded into boats and delivered to the furnace by means of overhead cranes.

The furnaces are of the open-top type. The internal dimensions of the crucibles are 22 x 9 x 6 feet. The bottoms are composed of dead-burned magnesite mixed with pitch and rammed in place. The bottoms are air cooled. The mixed charge is fed in around the electrodes, which are buried in the charge to a depth of about three feet. The furnaces work almost entirely as resistance furnaces, there being little arcing. As the electrodes are gradually eroded they are slipped through the holders and new sections of electrode are screwed on at the top.

The carbon in the charge deoxidizes the manganese, iron, and a portion of the silica, which mingle, forming the alloy known as ferro-manganese. The remainder of the silica and all of the alumina and magnesia in the ore combine with the lime, forming as slag.

The metal and slag are tapped from the furnace at regular intervals into pots and cast in sand moulds. When cool the slag and metal separate readily. The metal is broken up and shipped to steel plants, where it finds use in the deoxidizing of steel and in the manufacture of manganese steel.

Electric Power

In the earlier years of the Great Falls works all electric power used was developed on the smelter premises from the Black Eagle Falls and dam. At times the power so developed amounted to 8,000 or 9,000 horse-power. Now almost all of the electric power used is obtained from the power plants at Rainbow Falls and Great Falls of the Missouri River, which plants are situated a few miles down the river from the smelter and are owned by The Montana Power Company. About 1,000 horse-power is still developed on the smelter grounds from Black Eagle Falls and dam.

Electric power is obtained from The Montana Power Company over two 110,000-volt, and four 6,600-volt transmission lines. This energy is transmitted to four substations, namely, zinc plant substations, electrolytic substation, smelter substation and ferro-manganese substation.

The total capacity of the zinc plant substation is 40,200 kilowatts, or 54,000 horse-power.

The total capacity of the electrolytic substation is 8,200 kilowatts or 11,000 horse-power.

The total capacity of the smelter substation is 2,540 kilowatts or 3,400 horse-power.

The total capacity of the ferro-manganese substation is 24,000 kilowatts or 32,000 horse-power.

Desilverizing Kettles—International Lead Refining Company

Casting Refined Lead—International Lead Refining Company

THE REFINING OF LEAD BULLION

INTERNATIONAL LEAD REFINING COMPANY

GENERAL—The plant of the International Lead Refining Company is located at East Chicago, Indiana. The site has an extent of sixty-four acres, lying north of 151st Street, between McCook Avenue and the drainage canal. The Indiana Harbor Belt Railroad, Baltimore and Ohio, Chicago Terminal, and the Pennsylvania Railroad enter the yard at this plant. Construction started in April, 1912, and the plant was in operation six months later.

The primary purpose of the plant is to refine the lead bullion produced by the International Smelting Company, at it's Tooele plant at Tooele, Utah. In addition to this, the plant had been treating bullion from other parts of the United States, as well as from Mexico and Canada. Also a considerable amount of drosses and ores is purchased and treated at this plant.

OUTLINE OF PROCESS—The bullion, as it is received at the plant contains, in addition to gold and silver, certain impurities, which make the lead hard. These impurities are principally antimony, arsenic and copper, These have to be removed from the lead before it can be successfully treated in the subsequent processes. The furnaces in which the antimony and arsenic are removed are called "softening furnaces." The antimony and arsenic are removed as oxides, together with some lead oxides. The softened bullion is tapped from the softeners into large iron kettles where it is thoroughly mixed with a small amount of metallic zinc. The zinc comes to the surface and brings with it the silver and gold in the form of a crust. This crust is an alloy of zinc, lead, silver and gold that melts at a higher temperature than the lead. This crust is skimmed off into pneumatic presses which remove the excess of lead, leaving a dry alloy containing 3,000 ounces per ton of silver together with any gold.

This silver zinc lead alloy is treated in oil fired graphite crucibles or retorts and the zinc distilled off and condensed in graphite condensers. Most of the zinc is recovered and used again for desilverizing.

The metal which remains in the retorts is transferred to cupel furnaces which remove the lead, copper and other impurities and leaves the silver and gold. The lead is converted into litharge by means of an air blast on the surface of the bath and is skimmed off and retreated. The silver and gold (doré) is moulded into bars and shipped to the Raritan Copper Works for parting.

The lead in the kettles after removal of the bulk of the zinc is treated in a refining furnace under oxidizing conditions at a red heat. The remaining zinc, as well as any antimony or arsenic which may have escaped oxidation at the softening furnaces, is removed as oxide. From here the lead is run into the moulding furnace from which it is cast into a circle of moulds into bars of refined lead weighing about ninety-two pounds.

The skimming from the softening furnaces and litharge from the cupels, along with miscellaneous by-products are treated in a residue furnace, with coal and galena ore. The antimony concentrates into a slag, the silver and gold go into a lead bullion, and the copper goes into a matte. The lead bullion goes back to the softening furnaces for retreatment, while the copper matte is sold and shipped to some copper smelting works. The residue furnace slag, together with the zinc skimmings from the refining furnace, is treated in Blast Furnaces, the product being antimonial lead.

The Plant

SOFTENING FURNACES—The equipment consists of three softening furnaces, of a capacity of 220 tons each. Each softening furnace is built in a steel water-jacketed pan having inside dimensions 30 x 15 x 3 feet and 9 inches. They are coal fired. The bullion is charged directly from the railroad cars into the furnaces over a motor-driven charging machine. On the charging side of the furnace there are two sample kettles each of forty-five tons capacity. These kettles are used for melting down complete lots of bullion when

Refined Lead Bar—International Lead Refining Company

it is required to sample them. After sampling, the bullion is pumped into the softeners by means of a motor-driven centrifugal pump.

Desilverizing Kettles—On the opposite side of the softeners are six Desilverizing Kettles, two of 135 tons capacity each and four of 100 tons capacity each. Their chief use is to desilverize the softened bullion, and some of them may be used for sampling and miscellaneous melting and skimming operations. The pneumatic presses are raised and lowered into the kettles by means of jib cranes operated by compressed air at 100 pounds pressure.

Refining Furnaces—There is one refining furnace of 220 tons capacity built in a water jacketed pan of the inside dimensions 30 x 15 x 3 feet and 9 inches. It is coal fired. The lead is pumped from the desilverizing kettles by centrifugal pumps into this furnace.

Moulding Furnace—The "cooked" lead from the refining furnace runs by gravity to the moulding furnace. This furnace is built in a single pan of dimensions 24 x 15 x 4 feet and has a capacity of about 200 tons. The lead from the moulds is transferred by crane to the doors of the railroad cars for shipment or put into stock.

Residue Furnace—This furnace is built in a water-jacketed pan of dimensions 39 feet 3 inches by 10 feet 7½ inches by 3 feet 5 inches, it has a capacity of about 180 tons and will treat 70 tons of material per day.

Blast Furnace—There are two blast furnaces of circular water-jacketed type, 42 inches at the tuyeres with 5–3-inch tuyeres. Each furnace treats 50–60 tons of material per day. The slag from these furnaces is taken in large slag pots drawn by gasoline engine over industrial tracks to the slag dump.

Retorts and Cupels—There are ten retort furnaces fired with oil. They take a retort having a capacity of 1,300 pounds of charge. There are three cupel furnaces of Rhodes tilting type, having a capacity of five tons each.

Bag House—The gases from the blast furnaces go to a bag house of dimensions 50 x 65 x 50 feet, consisting of four chambers, each containing 144 woolen bags 18 inches in diameter and 30 feet long. The bags are shaken by an electrically-driven shaking device. The gases are taken from the Blast Furnaces through a long steel cooling flue of balloon type 7 x 9 feet and 2 inches and delivered to the Bag House by an eight-foot American Blower Company fan, driven by a thirty-five H. P. motor.

Stacks—There are two main stacks in the plant. One of them takes the gases from the Residue Furnace and one of the Softeners. The other stack takes the gases from the rest of the plant. The steam boiler has a metal stack of its own.

Power—The electric power used is furnished by the Northern Indiana Gas & Electric Company. It is delivered at 11,000 volts and transformed to 440 volts

ANACONDA WHITE LEAD

ANACONDA LEAD PRODUCTS COMPANY

In any treatise descriptive of the products of Anaconda Copper Mining Company or its numerous subsidiaries, the White Lead manufactured by Anaconda Lead Products Company must occupy its appropriate place as having within the past few years achieved a reputation and popularity in the trades which absorb white lead as a paint pigment, and for other purposes, which is only to be compared with the position of superiority accorded in the metal trades to the Electrolytic Copper and Electrolytic Zinc of the Company.

White lead is the oldest chemical pigment known to man, and until recent years the process of manufacture was not essentially different from that followed in ancient times. Numerous efforts have been made to modify it in the direction of quicker action and better control, but while the first aim has been attained to some extent, the latter has largely been missed, and the white leads produced by the several "quick" processes, while whiter than old-style white lead, are inferior to it in hiding power.

There remained, therefore, every incentive to discover and make practical some method of producing white lead which would remedy all these defects and give to the paint manufacturing and consuming public an ideal pigment. This has been accomplished in ANACONDA White Lead. The process of manufacture is an extremely rapid one, and every step of it is under mathematical control, so that the resultant product can be determined in advance as to its character.

The formula of white lead is $2PbCO_3.Pb(OH)_2$. This indicates a combination of carbonate and hydroxide of lead in the proportions of 69 to 31. It may be said that while much of the white lead produced by the "Old Dutch" and other processes may approximate these proportions, it is rather fortuitous that they do so, since there is absolutely no means of governing the formation; and they often vary to a considerable degree. It is only in Anaconda white lead that the process is under such control that if for any conceivable reason white lead should be desired of a different proportion it could be produced with ease.

The Anaconda process produces white lead of extreme fineness. It is amorphous, and is also of a soft texture, uniform throughout, and free from the hard, gritty, crystalline particles which characterize ordinary white lead. It is also entirely free from metallic particles, and from acetate of lead, and free from acid of any sort.

The first result of the process is a material of extreme whiteness. This in itself was enough to win instant favor in the trade, because a great deal of white lead is used for painting houses white, and naturally a pigment which was so much whiter than usual attracted immediate attention. The next point was that of economy, for it was found in practice that Anaconda white lead, when ground in oil and mixed for paint, was from 10% to 15% better in hiding power than old-style white lead, and, of course, better still than any "quick" process lead. This meant that a painter could obscure a given area with less lead, or, to state the case conversely, a given quantity of Anaconda white lead would cover more surface than any other white lead. This is due to its great fineness, which enables a painter to work more easily and quickly with it, and to brush it out thinner. As all paint experts are agreed that the thinner a coat of paint is, the harder it will dry, and the longer it will last, it will be seen that this is a valuable faculty of Anaconda white lead, besides the saving in material and time.

In addition to being used by itself as a paint, white lead is the fundamental and most important ingredient in ready-mixed paints, and the value of any such paint is directly contingent upon the proportion of white lead it contains, especially if it be destined for outside use. Anaconda white lead has showed its superiority in this situation. It mixes very readily with the other ingredients; its extreme fineness enables it to remain well in suspension, and as a pure white it does not "throw off" (or impart a dull cast to) delicate tints,—such as pink, light blue, etc.

The Printing Ink trade welcomed the advent of Anaconda white lead, for it provided them with a base of great opacity for white inks, and for other colors where extreme sharpness and great purity of tone are requisite.

Many other chemical pigments are made from white lead, and in the color-making trade, also, Anaconda white lead has proved its superiority, as it not only dissolves very readily by reason of its fineness, but its freedom from contaminations makes it almost totally soluble. The average percentage of insoluble matter is only .008%, which is far below that of ordinary white lead.

The plant of Anaconda Lead Products Company is located at East Chicago, Indiana; alongside that of International Lead Refining Company.

Spotted something that isn't quite right?

Let us know and we'll fix it for future reprints.
And as a thank you for your diligence, we'll
send you a free Sothis book of your choice.

publisher@sothis.press

This page intentionally left blank

Except of course for that line above. And this one.
And whilst we have your attention, perhaps you'd
like to follow us on Instagram @sothis.press. Go on.

Made in United States
Orlando, FL
04 November 2026

90204436R00037